Quantum Physics In Minutes

The Easy Guide in Plain Simple English for Beginners Who Flunked Math and Science

Donald B. Grey

Bluesource And Friends

This book is brought to you by Bluesource And Friends, a happy book publishing company.

Our motto is **"Happiness Within Pages"**

We promise to deliver amazing value to readers with our books.

We also appreciate honest book reviews from our readers.

Connect with us on our Facebook page www.facebook.com/bluesourceandfriends and stay tuned to our latest book promotions and free giveaways.

Quantum Physics In Minutes

Quantum Physics In Minutes

Table of Contents

Introduction 5

Chapter 1: Quantum Physics Basics 10

What is Quantum Physics? 10

Why Do We Need Quantum Physics? 14

The History of Quantum Physics 17

Quantum Physics Timeline 19

Chapter 2: Classical Physics and Quantum Mechanics 23

What Is the Difference Between Quantum Physics and Classical Physics? 23

Particles and Waves 33

Particle Wave Duality...................................34

Einstein and Quantum Physics 42

Essential Principles to Understand Quantum Mechanics 44

Chapter 3: The Quantum Fascination 48

Schrödinger's Cat and Equation.......................48

Superposition 51

Niels Bohr - The Quantum Theory 52

Quantum Entanglement...................................53

Quantum Physics In Minutes

Conformal Cyclic Cosmology 54

Quantum Physics Applications 55

Conclusion..63

Bluesource And Friends 69

Citations 70

Introduction

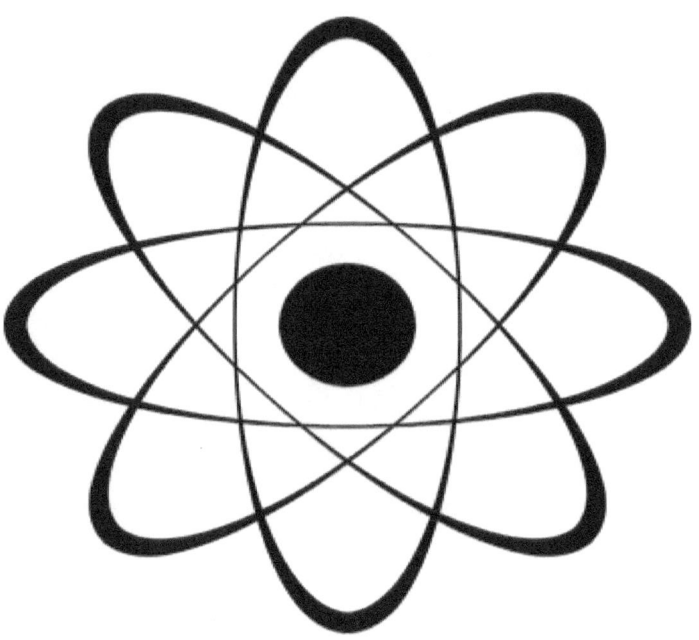

Figure 1.

Do you know of a discipline where it is taught that cats can be both dead and alive at the same time? Or a discipline where reality is dynamic and not absolute? 'Humanities' is probably the first word that pops into your head. The humanities *IS* where the creative writing arts department or philosophy department are, right? But what if I told you that you can find all of this in the sciences? Yes, in the sciences! Sometimes, truth is stranger than fiction. Are you intrigued? Would you like to learn more about these seemingly-impossible statements that I claim belong in the science department, where things are all logical and rational?

Then come and join me as we learn about quantum physics. Here, we will explore the more esoteric physics in the realm of science, and make it easier for you to understand. The field where these narratives fit in is called "quantum physics".

In a nutshell, quantum physics is the study of extremely tiny particles and how they interact with each other and their environment. This, in itself, is not what is interesting about quantum physics. We have dealt with atoms and molecules in chemistry, so dealing with objects on a molecular level isn't all that novel to you. Quantum physics is exciting because it's like a lawless cousin - it just doesn't quite follow the rules as we know it. It is seemingly an anomaly.

In quantum physics, particles and waves don't always act the way we have been taught they do. This can make things confusing. However, in this book, we're going to try to clear up the confusion so that you can have a good understanding of quantum physics.

Quantum Physics In Minutes

There is no need to feel alone - understanding quantum physics can feel like a mammoth task, but we're going to simplify it for you so that you can explore this fascinating topic. This book consists of an introduction, three chapters, and a conclusion. We will outline the following in the chapters to come:

- What is quantum physics?

    Here, we will look at the definition of quantum physics and explain its place within the science discipline.

- Why do we need to know about quantum physics? Does it have relevance for me?

    Here, we will illustrate the benefits of learning about quantum physics. Quantum physics will also be contextualized for you because it is not just for physicists.

- From where did quantum physics originate?

    This will include a history of quantum physics along with a few fun facts about the discovery of quantum physics.

- What is the difference between quantum physics and classical physics? Which one of the two is correct?

Quantum Physics In Minutes

- What is the relation between Einstein and quantum physics?

- Quantum physics theories:

   Schrödinger's cat will be explained

   Superposition

   Quantum theory

   Quantum entanglement will be illustrated

   Conformal cyclic cosmology will be touched on briefly

- Last but not least, we will look at the applications of quantum physics in our lives (as non-physicists or quantum physics specialists)

Aside from satiating your curiosity about the topic, you may wonder whether learning about quantum physics will help you in any other way? YES! Understanding quantum physics will provide you with the knowledge that will give you an edge in conversations, allow you to understand the intricacies of certain products, and allow you to appreciate the genius employed by science to create our cutting-edge technology. Are you excited? I know that I am! Let's get right into it.

Chapter 1: Quantum Physics Basics

This chapter aims to provide you with the basics of quantum physics and will answer three questions to get you all suited up for this journey into quantum physics. These questions include:

1. *What is quantum physics?*
2. *Why do we need quantum physics?*
3. *What is the history of quantum physics?*

These three questions will lay the foundation for your quantum physics journey. Let's go!

What is Quantum Physics?

The term "quantum physics" was coined by a group of physicists in Germany, which included physicists such as Max Born, Werner Heisenberg and Wolfgang

Quantum Physics In Minutes

Pauli at the University of Göttingen. The German term for quantum physics is *quantenmechanik*, which directly translates to "quantum mechanics". The terms "quantum mechanics" and "quantum physics" are used interchangeably in this book.

In the simplest terms, quantum physics is the study of very small particles that are generally called "quantum particles", although in other theories, they may go by the name "photons". To delve deeper into what quantum physics is, we need to understand that quantum physics is modern physics, while what we learn in school is classical physics. The difference between these two will be expanded on in Chapter Two.

Physics is the branch of science that is concerned with movement and the properties of things (Wikipedia contributors, 2020). *Quantum* is Latin for 'quantity,' and this is important because when we are dealing with quantum physics, we are dealing with discrete objects. After all, they follow the wave-particle duality (this will also be explained in Chapter Two). Thus, we can conclude that quantum physics is the study of the movement and properties of discrete objects. Put simply, quantum physics looks at microcomponents and attempts to explain how they

Quantum Physics In Minutes

interact with each other and why they interact in that manner.

From the time that we start school, we are learning about physics, intentionally or not. Long before we are introduced to the subject, we are introduced to the basics of physics. This includes learning about the push-and-pull forces, gravity, and the properties of materials. In school, we learn that atoms are the building blocks of matter. Atoms are quite small - they are not visible to the naked eye.

In quantum physics, everything that is observed, is at an atomic level or even smaller. According to Niels Bohr, the atom is made up of electrons, neutrons and protons. In quantum physics, we will be looking at how these particles interact with each other since they do not obey the laws of particles that we know and understand. It is important to note that quantum physics and classical physics differ significantly, and that will be explained with more depth in Chapter Two.

"No one really understands quantum physics" is a commonly-heard phrase, and while partially true, it is not all true. The reason quantum physics has this reputation is that, unlike other branches of science, it is not absolute, and calculations in quantum physics have an element of randomness and uncertainty. This

Quantum Physics In Minutes

is where the real fun in quantum physics begins, because there's always an adventure to be had when there's a little bit of uncertainty. There are no absolute solutions to quantum physics, because there are only probabilities which can be pretty daunting when you've only ever been exposed to classical physics.

If classical physics is the embodiment of order, then quantum physics is the embodiment of chaos to the ones learning it. Scientists actually do understand quantum physics, even though it boggles the mind. I would be very worried if they didn't understand quantum physics, and then suddenly decide to utilize it in nuclear power stations on *assumptions*.

Quantum mechanics makes use of four phenomena: Quantization of certain particles, quantum entanglement, the uncertainty principle, and wave-particle duality. These four phenomena will be explained in the book.

Why Do We Need Quantum Physics?

Are you reading this on your phone or your laptop? If you are, then in your hand, you are using a device that makes use of quantum physics. If not, then you probably have an LED light somewhere in your house. LED lights make use of the laws of quantum physics, so we can use energy more efficiently. There is also a high chance that the creation of your book used a device that makes use of quantum physics.

Quantum physics is the science that is going to revolutionize the way that we live. Understanding the basics of it will allow us to understand how it is being used. For scientists, understanding how particles react in different situations will give them the ability to maximize the potential usage of quantum physics.

You may already know this, but if you don't, then we are currently in the 4th industrial revolution where cyber-systems are the new 'IT' thing. Quantum physics has already played a part in the 3rd industrial revolution (which was characterized by electronic and information technology systems) by aiding in the development of faster computers. Thus, learning more about quantum physics (for both the scientist and the average person) will prove to be beneficial in the 4th industrial revolution, which will play a part in bettering both our computing technology and communication systems.

Quantum physics has long been sidelined due to the perceived difficulty of it. When you think of quantum physics, it is likely that you are thinking of highly-specialized experiments that can only be done in a lab, and have no direct effect on you. You're partly correct - the experiments are done in a lab, but their application does and will have a direct impact on you.

Quantum physics is what allows us to have faster and better computers, more accurate sensors and highly developed lasers. You can learn more about the various applications of quantum physics in Chapter Three. Due to the fact that quantum physics impacts us directly at home, it's worth learning about. Not only to understand how technology works, but also to strengthen your knowledge base.

There is another reason to learn about quantum physics, and it is more specific to those of us who are not studying quantum physics or anything that may relate to it. That is to prevent yourself from being scammed. Quantum physics is entering a 'second' revolution (The first one was in the early 1900s when it was introduced), and because of that, it is becoming increasingly popular.

However, it is still largely unknown, which means that there are people who will try to swindle you by trying to sell you ideas or apparatuses that 'use quantum physics.' Quantum physics is growing and soon, it will be widespread enough that it will be normal, so why not get a head start and learn all about it now? It also doesn't hurt to be knowledgeable about what's trending globally.

As you are aware, quantum physics is linked to our technological advancement; due to quantum physics, we have been able to use devices such as speedy computers, Blu-ray DVDs, various scanners, LED lighting systems, and many other things. The next part of this chapter will deal with a brief history of quantum physics.

The History of Quantum Physics

What is history but a storybook of past events? Quantum physics has a rich history that stretches all the way back to the 1800s, where scientists experienced quantum physics but did not know what they were dealing with. However, for the purposes of simplicity, we will only talk about the time from when quantum physics was discovered and named in 1900, until now.

The founding father of quantum physics is Max Planck. It is believed that Max Planck and Niels Bohr are the founding fathers of quantum physics, with

Quantum Physics In Minutes

Albert Einstein playing an important role in developing it further.

One thing that quantum physics revealed about the majority of physicists, is that they are stubborn and almost fanatical in their beliefs. Definitely not a bad friend to have behind you! However, in terms of quantum physics, it has taken well over 70 years for physicists to accept quantum physics as a 'real' science, because quantum physics took the carefully-delineated lines of classical physics (more on that in Chapter Two) and blurred them. The last attempt to refute quantum physics came in 1977! It took an astounding 70-odd years for physicists to come to terms with the fact that the world is not as clear cut as classical physics would have you believe.

Not only were the lines blurred between properties that are clearly outlined in classical physics, but quantum physics created a platform where the results were no longer as absolute when it comes to its own mathematical, scientific and philosophical solutions. To help you understand how quantum physics came into being, I would like to provide you with a timeline and explanation of the history of quantum physics. Buckle up, it's going to be a wild ride!

Before we start with the history of quantum physics, please note that there were a myriad of scientists who contributed to what we consider to be quantum physics today. Quantum physics as we know it today is the sum of contributions by all of these scientists mentioned below and more. Some of these people include Niels Bohr, Schrödinger, Heisenegger, Copenhagen, and Einstein. Together, their theories affect what we now know as quantum physics. Since its inception, quantum physics has played a pivotal role in the progression of technology.

Quantum Physics Timeline

- In 1900, a German physicist named Max Karl Ernst Ludwig Planck accidentally became a revolutionary and the founding father of quantum physics at the age of forty-two. Max Planck was attempting to explain the spectrum of light. Planck knew that the spectrum of light should have a specific shape, but while he had the mathematical formula, he had no theoretical justification for his formula.

  In an act of desperation, he resorted to using a mathematical trick which showed that, instead of objects having infinite energy, objects had 'oscillators' that emitted a discrete amount of energy. Planck did not intentionally mean to create quantum physics - it was a byproduct of his need to fulfil a goal, which was to derive an expression for the spectrum of electromagnetic energy that is released as radiation when an object is heated. This will be elaborated on more in Chapter Two, when we explain the theories in quantum physics.

- Between the years 1903 and 1905, Albert Einstein used Planck's theory to help him with the explanation of his photoelectric effect, which later won him a Nobel Prize. Einstein explained that light was a wave that was made up of many particles (similar to Planck's 'oscillators'), and these particles were called "photons."

- In 1923, Arthur Holly Compton, a scientist, conducted an experiment with an x-ray. In this experiment, he concluded that waves act like particles when bounced off materials, which again complies with what Planck and Einstein stated in previous years. So between 1900 and 1923, the phenomena of waves acting like particles was the rage (or scourge depending on your perspective) of the physics world, but then, along came Broglie.

- In 1923, Louis Victor Pierre Raymond de Broglie, a French scientist, had the thought that there should be symmetry or balance in the world, so if waves could act like particles, then it stood to reason that particles could also behave like waves. In 1927, de Broglie's theory was proven, and in 1929, he received a Nobel Prize for the prediction of the wave nature of particles. Then in 1937, Davisson and Thomson received a Nobel Prize for proving the wave nature of particles. So one of the core principles of quantum physics was laid - the particle-wave duality.

- In 1926, Schrödinger created his equation, which was a mathematical expression of what wave functions look like. He also created the popular paradox of Schrödinger's cat, where the cat lies in a state of superposition until observed or measured - meaning that it is both dead and alive until measured or observed.

So far in this chapter, we have looked at what quantum physics is, why we need it and what its history is. This should give you a good understanding with which to go forward. The next chapter is going to get a little more technical than this chapter, as we are going to look at the difference between classical physics and quantum physics, as well as the various theories that influenced each field of physics.

Chapter 2: Classical Physics and Quantum Mechanics

Now that you have been introduced to the basics of quantum physics, it's time to understand where it fits in the science world. Physics has been divided into two parts: Quantum physics (or quantum mechanics) and classical physics. The reasons for this separation will be discussed in this chapter so that you can understand what sets the two streams apart, and who the popular proponents are within each.

What Is the Difference Between Quantum Physics and Classical Physics?

Physics is the study of the interaction between things along with their properties. The very first difference between classical physics and quantum physics is that their focus is different. Classical physics focuses on the world that we can see with our naked eye, which is the macroworld, and quantum physics focuses on the world that we cannot see with our naked eye, which is the microworld.

An easy way to distinguish between classical physics and quantum physics is to know when the laws and principles were conceived. The laws and principles of classical physics are those that predate the 1900s, while those of quantum physics came after the 1900s. The second major difference is that they seem to contradict one another. They are at seemingly opposing ends of the spectrum that is physics.

Classical physics includes Newtonian mechanics, Maxwell's theory of electromagnetic field and Einstein's theory of general relativity (Cresser, 2011). These theories are believed to portray what is really happening in the physical system that we are observing. Classical physics is based on causal effect, which is simply a chain of events where event B is influenced by event A.

The problem with classical physics is that the laws and principles crumble when you attempt to apply them to very small things. As such, classical physics is best used when dealing with the macroworld. Classical physics is the physics we learn in school. It comprises topics such as electrostatics, electric circuits, movement, motion and sound, to name a few. Below are a few of the proponents of classical physics.

Isaac Newton - Newtonian Mechanics

Isaac Newton was born on December 25, 1642, and he can be considered the founding father of classical physics as we know it (Filmer, 2013). The laws and principles that he discovered regarding motion and gravity are what propels classical physics (Encyclopedia.com, 2020), and although Newton is known for many discoveries, he is known best for his theory of gravity, along with his three laws of motion.

Three laws of motion:

1. First law: The law of inertia - this simply states that an object will remain in its current state unless a force acts on it. Below is an example. If you pulled the yellow tablecloth fast enough, and the tablecloth did not exert a lot of friction (force) on the bowls, the bowls will not move.

2. Second law: The rate of acceleration is directly proportional to the mass and force of an object.

Quantum Physics In Minutes

3. Third law: For every action there is an equal reaction. When you place your hand on the frame, you exert a force onto the frame. Similarly, the frame exerts an equal but opposite force onto your hand. That neither your hand nor the textbook moves shows that the forces exerted equal each other out. If more force is exerted from either end, movement may occur.

These 3 laws form part of the foundation of classical physics.

James Clerk Maxwell - Maxwell's Theory of Electromagnetic Field

Maxwell's theory has four equations which factor into the electrical and magnetic topics in classical physics. Maxwell's theories show that electrical and magnetic forces are not separate, but are demonstrations of the same thing: Electromagnetic forces. I will briefly outline them below.

1. Electric field charges originate on positive charges and end on negative charges.
2. Magnetic fields are continuous and have no end.

3. A changing magnetic field produces an electromotive force, which creates an electric field.
4. Magnetic fields are generated by changing electric fields.

Albert Einstein - General Relativity Theory

Albert Einstein was a German-born physicist who contributed massively to both classical physics and quantum physics. One of Einstein's principles that classical physics follows is general relativity. This theory was developed in 1915 (one of the few classical physics exceptions), while the theory of special relativity (which influences quantum physics) was developed in 1905.

Einstein's theory of general relativity is about space and time and refers to the effects on spacetime with factors such as matter, gravitational forces, energy and momentum.

These are three of the biggest proponents in classical physics.

Classical physics is quite clearly demarcated because everything has its place and follows a rule. However, when it comes to quantum physics, we can see that this is not true for the microworld. In the following

Quantum Physics In Minutes

pages, I will provide a brief list of the proponents of quantum physics, and then I will provide a more detailed explanation of particles and waves, as they are the reason for the divergence from classical physics.

Quantum physics includes theories from Max Planck, Albert Einstein, Arthur Holly Compton, Louis de Broglie and Erwin Schrödinger. The advent of quantum physics introduced a few limitations to classical physics, such as blackbody radiation.

According to classical physics, objects that are heated will radiate all of their energy in electromagnetic waves. If this is true, that means that particles have infinite energy. However, as we have ascertained, Max Planck disproved this when he showed that light particles have a certain amount of energy. This is also corroborated by Niels Bohrs who, through quantum theory, explains that electrons have specific energy and can only be in specific orbits.

Below, we will look at the theories by Planck, Einstein, Compton, de Broglie and Schrödinger.

Max Planck

Max Planck was a German physicist who unwittingly discovered quantum physics. Planck needed to find the answer to the light spectrum, so he worked backwards by looking at the effects and tracing the cause. He had a mathematical formula but needed theoretical justification for it.

In order to make his theory work, he added tiny 'oscillating' particles, with the intention of mathematically removing them at a later stage once his theory was proven. However, to his surprise, the formula worked, but it only made sense if h had a specific value that was not zero. He then explained that the energy of the oscillator is determined by its frequency. Planck's formula to find the energy of Energy is:

$$E = hf$$

Where h = 6.62607004 x 10^{-34} m² kg/s, f = frequency and E = energy

Planck thus solved the ultraviolet catastrophe (or the Rayleigh-Jean experiment) that was done a few years prior. The Rayleigh-Jean experiment stated the radiation emitted would be in all frequency ranges, but that as the frequency increased, so too would the emission of energy. This turned out to be false because it assumed that particles have an infinite source of energy. What Planck showed was that particles have a discrete amount of energy.

Quantum Physics In Minutes

However, by showing that light is not a continuous wave, but rather exists as individual particles, Planck created an upheaval in the scientific community, because waves are supposed to act like waves, and waves do not have particles because they are continuous disturbances. And so began the advent of quantum physics. Planck did not intend on creating quantum physics, but it was a byproduct of what he wanted to prove.

Albert Einstein

Albert Einstein is arguably one of the greatest pioneers in the science discipline. He has theories that support quantum physics and classical physics. There are two things that I want to bring to your attention regarding Einstein's contribution to quantum physics. The first is the photoelectric effect, and the second is special relativity.

1. The photoelectric effect occurs when you shine a light onto a material, and it emits electrons (Wikipedia contributors, 2020). Below is an illustration of the photoelectric effect. As you can see, the light waves hit the solid metal plate, and electrons are ejected.

Quantum Physics In Minutes

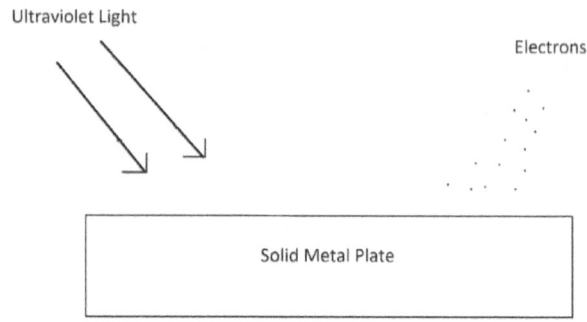

Figure 1.

To provide a mathematical representation of this, Einstein used the formula:

$$p=h/y$$

where p = the photon, h = Planck's constant, and y = the wavelength of the light.

This theory formed the beginning of technologies such as lasers. But it also created a disturbance in the physics community, because it now seemed to provide further proof that light, previously known as a wave, was actually made out of particles. So which law did it follow? More on this in Chapter Three.

Quantum Physics In Minutes

Scientists such as Robert Milikan tried to discredit Einstein's photoelectric effect, but he only managed to confirm it (Orzel, 2009). Even though Milikan's experiments proved that Einstein's theory was correct, it was not enough to sway physicists. (I did say they're a stubborn bunch!) It was in 1923 when Einstein's photoelectric theory gained prominence, because of the experiments of a scientist named Arthur Holly Compton.

2. Special Relativity

   The theory of special relativity explains how space and time are linked, for objects that are moving at a consistent speed (Howell, 2017). This differs from general theory of relativity because the theory of special relativity applies in the absence of gravity, while the theory of relativity needs the laws of gravity to explain its effects on the system.

Arthur Compton - Proving Waves' Nature as Particles

Arthur Holly Compton is the physicist who proved that waves indeed act like particles. He did this in an experiment called the Compton effect in 1923, for which he won the Nobel Prize in 1927. The Compton effect was demonstrated by pointing an x-ray at solid material. It was noted that the collision of the x-ray waves with the solid produced a reaction that was the same as two particles colliding.

Particles and Waves

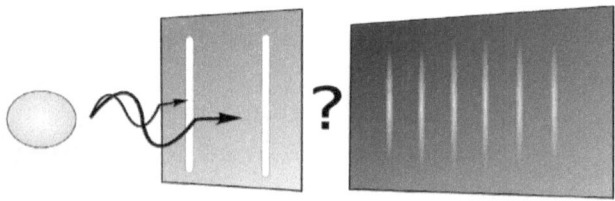

Figure 2.

There is a fundamental difference in the way that particles and waves are thought about when looking at the difference between quantum physics and classical physics. We will start by looking at the definition of particles and waves in the classical physics sense.

In classical physics, particles and waves are distinct entities that have separate defining characteristics. A particle is characterized by its location and mass, while a wave is characterized by its frequency. The two also look very different.

Particle Wave Duality

According to classical physics, particles behave the way that you and I are taught in school - they have weight, location and speed. If you have all of these aspects, then you can calculate its momentum, and if you can understand the particle's momentum, then you will understand how the two particles interact with each other when they collide.

For instance, if the momentum of particle A is a lot bigger than particle B, then when particle A collides with particle B, both of them will roll. Particle B will roll because particle A has "pushed it," and particle A will continue to move because particle B's mass was too small to stop it. This then follows Newton's laws of motion. According to classical physics, the number of particles can also be counted.

Again, according to classical physics, waves, contrary to particles, are a moving disturbance that is spread out in a sort of pattern that moves over time. There is usually no physical object that moves when dealing with waves, but you can see the disturbance produced by waves. An example of a wave that manifests in a physical disturbance, and that we can see in the real world, is a ripple in the water when it is disturbed by a drop of water.

Figure 3.

Waves do not have a specific position, so you cannot locate them, but you can follow the pattern of a wave if you can look at a part of the wave. Unlike particles, waves cannot be counted because they are continuous, but you can find out the frequency of waves. You are also unable to add waves in the same manner that you add particles, because waves are not discrete but are continuous. Additionally, they have different properties, so trying to add them the same

Quantum Physics In Minutes

way is akin to trying to make bread with oranges the same way you would make bread with flour.

In everyday life, we experience waves in terms of light and sound, but these waves differ significantly. The main reason is because of how they act. Sound waves are long, and they can bend, while light waves are short, and they move in a straight line.

Seems like pretty simple and standard stuff, yes? Everything is clearly categorized and each has its own characteristics in classical physics. But then a German physicist named Max Planck came along, and he was studying something called "thermal radiation". Thermal radiation is radiation that can be seen. An example of radiation includes infrared light. Thermal radiation is caused by heating up an object which will then emit light radiation that is visible.

When Planck was studying thermal radiation, he was trying to solve for the spectrum of light emitted. The reason for this is because of the ultraviolet catastrophe. The ultraviolet catastrophe, also known as the Rayleigh-Jean experiment, occurred just before the 20th century started, and the premise of this experiment was that a blackbody would emit radiation with infinite power. It was presumed that the higher the frequency, the more powerful the radiation would be, but this did not prove true in the experiment.

Quantum Physics In Minutes

Thus, when Planck was trying to establish the spectrum of light, he knew that there was a specific spectrum, and he even had a formula for it, but there was no theoretical justification. How can it be (classical) physics if there is no justification? So he resorted to using a mathematical trick to prove his theory with the intention to simply remove the additions at a later stage, once he proved his theory.

However, it backfired! How, you ask? Well, in order for his theory to work, he added in a mathematical representation of 'oscillators', but in doing so, he showed that light can act like particles, and unintentionally upset years of 'fact!'

Classical physics dictates that waves and particles act differently, but with Planck's theory, light (waves) can act like particles and waves, which just throws everything that we just spoke about in the previous paragraph out of the window! An accident turned out to be the start of quantum physics as we know it.

Planck's theory was furthered by Albert Einstein when he used Planck's theory to explain the photoelectric effect. According to Einstein, light consisted of a little stream of photons, and each one had energy that can be provided if you have the frequency. This corroborated Planck's theory of waves acting like particles.

Quantum Physics In Minutes

Furthermore, the relativity theory by Einstein states that photons have momentum even though they lack mass, and this is because the photon has energy. This means that the collision of a wave and an object should look and act the same way a collision would if it were between 2 particles, which then follows Newtonian Laws. Photons with a small wavelength will have lots of momentum while photons with a large wavelength will have little momentum, and the momentum will impact the force and acceleration of the objects.

This theory of Einstein's that was built on Planck's was not accepted until about 1923, when a scientist named Compton did an experiment that proved what Einstein was saying. Thus, we came to find out that waves can act like particles, contrary to the laws and principles of classical physics.

In 1923, a physicist, Louis Victor Pierre Raymond de Broglie, suggested that if waves could behave like particles, then why shouldn't particles behave like waves? So he decided that he would show that a symmetry exists. Broglie suggested that particles should have a wavelength that is determined by its momentum, similar to a photon's momentum being determined by its wavelength, so he inverted Einstein's formula.

Quantum Physics In Minutes

Understanding the duality of waves and particles gave birth to a third type of particle, called the quantum particle. To provide a visualization of this particle, you can imagine it as a coin, where the waves and particles, as described in classical physics, are the sides of the coin. Previously, in classical physics, we thought that waves and particles were two separate entities, but through quantum physics we now find that they are one entity that can behave in different manners.

Figure 4.

Wave-particle duality is one of the fundamental tenets of quantum physics because it shows that physics is not as binary as we thought. Finding out about the wave-particle duality does not negate all the laws and principles that we know from Newton, but rather it provides us with a new perspective. Newtonian laws do not impact quantum physics laws and vice versa. This is because it is extremely difficult to see quantum laws since we are dealing with such minuscule particles, where we can easily see a Newtonian law in action, because they are macroscopic.

Quantum physics allows scientists to understand the atomic structure of materials, which allows them to better use them. This has helped in creating stronger and lighter materials for cars, planes, and cell phones, among other things.

Figure 5.

Einstein and Quantum Physics

Albert Einstein had an important relationship with quantum physics, because he not only believed that waves could be made of particles, but he also endorsed that particles could behave like waves. If he had never done that, it is likely that we would have not been introduced to quantum physics as we know it now. It might have taken a lot longer to verify the authenticity of quantum physics.

Although Einstein was one of the key contributing physicists to quantum physics, he was also vehemently against it for philosophical reasons. Einstein believed that the world was deterministic, meaning that he believed that the world is governed by cause and effect that can be clearly demarcated and measured, much similar to the principles that we are taught in classical physics. In classical physics, there is no room for interpretation - there is only the reading of results that are simple and clear.

However, in quantum physics, there is a lot of interpretation that is needed, so it is sometimes said that no one really *knows* what is going on in quantum physics. One of Einstein's great quotes for refuting the randomness of quantum physics included him saying that he believed that quantum physics was missing something.

"The more success the quantum theory has, the sillier it looks"

He tried to show this by teaming up with other physicists in the 1960s and doing an experiment called the "EPR". EPR is short for the "Einstein-Podolsky-Rosen thought experiment". This thought experiment sought to prove that quantum physics was missing a crucial part. Unfortunately for Einstein, this thought experiment did not refute quantum physics as much as he thought it would, and actually aligned with it.

The EPR claimed that the measurement of one particle, in an entangled pair of particles, would have an instantaneous effect on the other particle, regardless of the distance between the two. This would indicate some sort of communication between the two particles that is faster than light, which is not the case. That would violate the notion of relativity, which states that information cannot travel faster than light.

Einstein was not trying to disprove quantum physics, but he was trying to grapple with his internal self that could not believe in the randomness of quantum physics. This can be captured in Einstein's famous quote about a higher being not gambling with life by creating something ambiguous:

"God does not play with dice."

Naturally, one can see the existential crisis that might arise if you believe that everything is utterly random and that there is no causal effect in place. This, of course, does not exclude that a sense of cognitive dissonance could also be at play because of the nature of classical physics.

Essential Principles to Understand Quantum Mechanics

- **Wave Functions**

  Everything in the universe can be described by a quantum wave function (sometimes written as "wavefunctions"). A wave function is a mathematical formula that governs the behavior of wave functions. Schrödinger's equation is the mathematical formula that governs quantum wave functions. Given the basic information, one can calculate quantum

Quantum Physics In Minutes

wave functions by using the Schrödinger's equation, in the same way that you can use Newtonian laws to calculate the mass or location of an object. In this regard, Schrödinger's equation determines the observable properties of the object.

- **Allowed States**

  A quantum object can be observed in only ONE of the limited number of allowed states. In terms of light, one photon is one quantum of light, and that quantum of light can never be split. Therefore, there will only be whole numbers when dealing with quanta.

  An electron orbiting the nucleus of the atom can only be found in very specific states. Thus, each state has a particular energy and the electron will always be found with one of those energies, but it will never be in between. The electrons will then move between energy levels. The movement between energy levels is what is the origin of the term, *quantum leap*.

- **Probability**

  The wave function determines the probabilities of different allowed states. This allows philosophical concerns to creep in because there is no absolute answer, which leads us to wonder about higher powers and religion. These concerns are brought about because quantum physics allows you to calculate probabilities but not absolute certainties. It is one of the most significant differences between classical physics and quantum physics.

- **Measurement**

Measurement in quantum physics is an active process, meaning that it is constantly occurring because it is constantly being changed. The act of measuring something quantifies the reality that we observe, which then implies that if we do not measure it, then there is no reality.

To describe Schrödinger's cat theory, we look at measurement because we need to use a wave function that has two parts - one part for the cat to be alive, and one part for the cat to be dead. When we have a 2-part wave function, it does not mean that the object is in one of two states, but that it is in both states at once - this is called superposition.

Quantum Physics In Minutes

In conclusion of this chapter, we can ascertain that quantum physics and classical physics are different because of their focus and the causal-versus-random characteristics that each displays. However, quantum physics is not to be disregarded. Even though it is random, it has been tested and proved extensively, and influences the building and use of things such as nuclear energy and LEDs.

We reviewed the laws and principles from before 1900 that are considered classical physics, and those that came after, that belong to modern physics, including quantum physics. Although quantum physics is relatively new (it is only about a century old while classical physics boasts a grand three centuries), it is quite advanced and promises to lead us to a second quantum revolution.

It is also worth noting that the laws and principles of classical physics are not rendered null and void by the advent of quantum physics, or modern physics, because classical physics is still an accurate depiction of the reality in the macroworld.

Last but not least, we looked at the four principles you need to know in order to understand quantum physics. In the next chapter, we will go further in depth so that you can understand the more fascinating aspects of quantum physics.

Chapter 3: The Quantum Fascination

There is a fascination with quantum physics because of its esoteric nature. It has a reputation for being difficult to understand and having principles that conflict, but I am going to explain these principles so that you have a firm grasp on the basics of the quantum physics theory. These principles are explained in simple English with very little math, unless absolutely needed. The scientific terms are also explained so that you do not get lost on the way.

Schrödinger's Cat and Equation

Erwin Schrödinger succeeded Max Planck, and he provided two important things to quantum physics: The thought experiment called "Schrödinger's cat", and Schrödinger's equation, which is used to describe the wave function in quantum physics. Erwin Schrödinger, who was born in 1887, was an Austrian-

Quantum Physics In Minutes

Irish physicist and Nobel Prize winner. Let's start by looking in greater detail at one of the most famous experiments that is linked to quantum physics.

Schrödinger's cat is one of the more infamous theories that belong to quantum physics. Please note that a cat was never harmed, because this was not a practical experiment but rather a thought experiment. The premise of a thought experiment is to create a well-defined and fully thought-out hypothetical scenario. This particular thought experiment was created by a physicist named Erwin Schrödinger.

Schrödinger's cat can be explained as follows: First, a cat is put into a box. Along with the cat is a radioactive mass and poison. The box is then sealed and left for a period of time. During this time, there are two possible outcomes that are happening at the same time - the cat is dead and alive.

This is called "superposition", and will be elaborated on later in this chapter. When we open the box to inspect, one of two outcomes will present itself - the cat will either be dead or alive. No matter how many times you repeat this process, you will always find the cat in one of these two states. Look at the image below to see a representation of Schrödinger's cat.

Figure 6.

Schrödinger's equation is an equation that describes the wave function in a quantum mechanic system. In classical physics, there is a predicted path provided where you have all the variables. However, this is not the case with quantum physics because of things like the uncertainty principle. Thus, Schrödinger's equation can be said to be the equivalent of Newton's second law in quantum physics.

Schrödinger's equation is

$$\frac{h}{i}\frac{d}{dt}|\Psi> = \hat{H}|\Psi>$$

Quantum Physics In Minutes

In a nutshell, here is what you are calculating when you use Schrödinger's equation:

$$i = \sqrt{-1}$$

$$\underline{h} = \frac{h}{(2 \times \pi)} \text{ where } h = 6.62607004 \times 10^{-34}$$

$$\frac{d}{dt} = \frac{Distance}{time}$$

$|\Psi>$ = the quantum state

$\hat{H}$ = Hamiltonian operator - this is the sum of all the different energies and how they affect the quantum state

We are not going to get into the exact math of Schrödinger's equation, but basically Schrödinger's equation shows how the quantum state changes with time depending on the energies present, which provides us with a mathematical expression of how a quantum system works.

Superposition

Quantum Physics In Minutes

Superposition in quantum physics is when a particle exists in two states at once. In quantum physics, superposition is the state in which particles exist until an external force, such as measurement, acts upon the system. Once there is an external interference, then the superposition state collapses, and you end up with one of the possible states. To illustrate this, you can use Schrödinger's cat as an example. Before you open the box to see whether the cat is dead or alive, the cat exists in a superposition state where it is both dead and alive, and only when you observe it (the interference in this case) will a single possibility between the two be made.

The concept is similar in quantum physics except that we are not necessarily discussing waves, but rather states, when we are looking at quantum physics. A good illustration of superposition states in quantum physics can be explained by Erwin Schrödinger's thought experiment where the cat in a box is both dead and alive until the box is opened. Thus, the cat is in superposition until you try to observe or measure it.

Niels Bohr - The Quantum Theory

Quantum Physics In Minutes

Niels Bohr is a Danish physicist whom you may recognize from your notes at school. He is the scientist that gave us the atomic model, where the electrons orbit around the nucleus of the cell. Although this model may be the reason many of us recognize his name right now, Niels Bohr is also known for something else - quantum theory! Quantum theory proposes that electrons orbit the nucleus of the atomic cell, but are limited to the orbits they can be in, as well as the energy that they lose as radiation, when they move from one orbit to another. Bohr's quantum theory explains why atoms release light that has fixed wavelengths.

Quantum Entanglement

Quantum entanglement is when the state of a particle in a quantum system cannot be described independently of the system. If the particle can be described independently of the system, then it is not entangled, but rather separable. To illustrate this, imagine that we are dealing with two electrons where the distance is negligible. These electrons can either spin up or down. If we decided to measure the spin of the molecules, we would find one of four possible combinations. If we did not measure or observe the particles, then they would be suspended in a state of superposition.

However, suppose that we assume that the particles are entangled (without getting into the nitty-gritty math), then the two particles would be in a relationship such that observing or measuring one particle would affect the other one. What do I mean by this? If, in an entangled system, you look at electron A, then you would immediately know the state of electron B, because they are entangled.

What's crazy about this is that there is no communication, per se, between the two electrons. It is almost as if each electron 'knows' when you exert a force on the other, and thus assumes the necessary state once the external force is exerted. Quantum entanglement is believed to be a resource for quantum computing (Johnston, 2018), which is super exciting because that means we will have more powerful computing devices.

Conformal Cyclic Cosmology

Quantum Physics In Minutes

This theory essentially states that the world does not have a beginning because it is in a perpetual state of 'reincarnation,' where each reincarnation is characterized by a cataclysmic event such as the Big Bang (Hindu, 2017).

Quantum Physics Applications

There are many different ways in which quantum physics is used in the real world. The very device that you are using to read this book probably has a camera; if it does, quantum physics has been used in its creation. Quantum physics is responsible for technology, such as nuclear power stations, MRI scans, laser beams, computers, toasters and many other smart devices.

These devices all make use of quantum physics knowledge in order to work. Your toaster also makes use of quantum physics when the element is heated, because millions of atoms vibrate and end up emitting packets of energy, which we see as the red light of the grill in your toaster.

Quantum Physics In Minutes

For instance, your mobile device contains many components, such as transistors, which all use quantum technology (Quantum Technology, n.d.). Your phone is not yet a quantum computer, but the silicone that it uses is inspired by the quantum mechanics behind the silicone.

Another application of quantum physics that is quickly gaining momentum is the advent of LED lights. These lights provide bright light at a fraction of the electricity cost; they don't need to be heated nor do they release a lot of heat, so they are durable and easy to use.

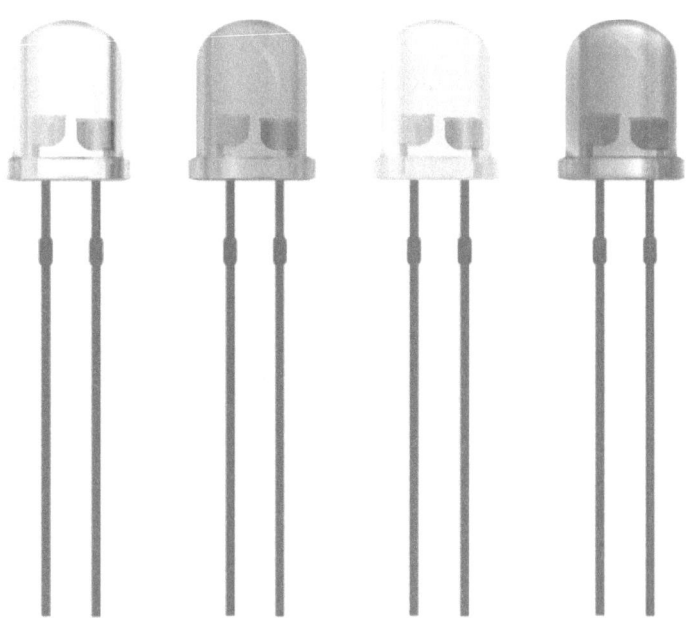

Figure 7.

Technology is also going to expand because of quantum physics. Some notable projects include a quantum computer, as well as quantum cryptography systems. This will allow for speedy encryption of messages that would be difficult to decrypt if someone hacks your account. Quantum computing is when you use phenomena, such as quantum entanglement and superposition, to perform complex mathematical calculations.

The advancement of quantum computing is important because it works much faster than classical computers. Quantum simulation may also hold the key to understanding and bettering nanotechnology. Because it will simulate the behavior of the atoms and particles, it can be better understood and utilized.

Figure 8.

There are also many uses of the quantum physics application in laser communication. You can see this whenever you utilize the internet via your Wi-Fi connection. There are little packets of information that are sent back and forth between your devices. It can also be seen in the cables itself, which acts as a carrier.

Quantum Physics In Minutes

Figure 9.

Another reason to learn about quantum physics is so that you do not get scammed. Right now, quantum physics is gaining a lot more traction than it previously did (I mean, it took a little over 70 years for physicists to accept quantum physics as a real science). So, unfortunately, there are many people who are misinterpreting quantum physics - either intentionally or unintentionally. Those who are deliberately misinterpreting it can give you information that is quite misleading.

Currently, there are people who are explaining quantum physics as this 'magical' science that will help enhance your life, but the truth is that quantum physics can only be seen in application or in a lab. You cannot see it with the naked eye, nor will you be able to see the interactions without isolating the particles and creating a stable environment in which to observe these particles.

Included in this 'magical' science, that people may try to tell you about, will be the ability to create infinite amounts of energy by harnessing quantum physics. Unless you are a physicist, it is unlikely that you will be able to create energy using quantum physics, and even in quantum physics, there are limitations, as explained in Planck's discovery where each oscillation had a specific amount of energy that it could hold.

Figure 10.

Quantum Physics In Minutes

Although it is terrible that people want to scam others, I agree with Chad Orzel when he states that it's a good thing that people are trying to scam others with quantum physics. It means that people are seeing the value of quantum physics and trying to get ahead (kind of like you reading this book).

History has proven that when something is new and innovative, people will use it to scam others. An example of this is electricity. Before it was widespread, people would try to sell products that used electricity, by saying that it was cutting-edge technology and could add significant value to your life.

The other 'magic' that they will try to sell you had to do with your health and well-being. Quantum physics may help you if you are in a hospital or in need of some kind of equipment that makes use of quantum physics technology. But it will not heal you or take away your depression if you 'channel' the quantum energy of your mind to influence your body.

Hopefully, you will be able to spot the scammers ahead of time and turn the other way, because you will now have a good grasp of what quantum physics entails and how it works.

Quantum Physics In Minutes

To conclude this chapter, I will summarize what we have gone over. We have looked at Schrödinger's cat and his equation. We then further looked at superposition, which was utilized in Schrödinger's cat paradox. We also looked at Bohr's Quantum theory and how it influenced quantum physics. Then we looked at quantum entanglement - one of the many weird but wonderful parts of quantum physics.

Conclusion

Thank you so much for reading this book! I am so proud of you for making it this far. You are doing great! I hope that you have enjoyed reading it as much as I have enjoyed writing it for you. I wish you success and fun times in your future quantum physics adventures. In this book, we have discussed many topics, and I hope that they will help you on your journey to further understanding quantum physics.

Quantum physics is a wonderful and exciting topic that is promising to revolutionize the world as we know it through advancement of technologies that improve our quality of life. So far, we have understood that quantum physics is the study of the microworld and it is a topic that is exciting to be well-versed in.

We have simplified it for you so that you understand what it is about. We have also looked at the history of quantum physics to give you an idea of how quantum physics came to be and who the founding fathers are. We then delved into the differences between quantum physics and classical physics.

Quantum Physics In Minutes

The main differences are that classical physics has clearly demarcated rules and laws that are used to govern the macro world - that is the world that we see with our naked eye, while quantum physics, the physics that deals with the world at a micro level, the level we cannot see with our naked eye, has contradictory laws to classical physics. This created a huge upheaval in the early 1900s when it was discovered.

Even now, it can be very confusing to understand when classical physics comes into play and when quantum physics does. We live in the macro world, so for the most part, we follow the laws of classical physics.

The various proponents of classical physics are Newton, Maxwell and Einstein. Their contributions to classical physics are explained as well as their theories. We then further delved into the proponents of quantum physics and provided an overview of their theories.

The laws of classical physics conform to the Newtonian laws, while quantum physics conforms to four basic principles, wave functions, allowed states, measurements and probability. These four basic principles were elaborated on, and we then discussed the various theories offered through the years on quantum physics that explain this fascinating topic.

For example, we looked at Schrödinger's cat. This was a thought experiment, so no cat was actually harmed, but the experiment did show us that the cat can be both dead and alive until we observe it, at which point we will know about the status of the cat. We also look at Schrödinger's equation, which provides the mathematical expression of what should happen in a 'physical' quantum system. Then we explained superposition, which is the act of two things coinciding in a state of multiplicity. When superposition collapses, you are seeing one state.

Next up we explained quantum entanglement - a phenomenon that is characterized by particles in the same allowed states, where the act of measuring one will give you a measurement of the other. This measurement is not necessarily the same. It is simply a state of being. For instance, if you had two Schrödinger's cats, then you would be able to say that in one box, the cat is alive, and in the other, the cat is dead.

Quantum entanglement dictates that once you open one of the boxes and find the state of the cat, you will immediately know the state of the other cat. Obviously this is to do with particles and not cats because quantum physics works at a micro level, and it is very, very difficult to observe at a macro level. We also briefly looked at Conformal Cyclic Cosmology, which simply states that the Earth is in a loop of 'reincarnation'. Last but not least, we looked at Niels Bohr's quantum theory, which states that electrons move around, but they can only move around in prescribed orbits.

If you would like to do more reading, then I suggest you look at a few books. The first is, *How to Teach Quantum Physics to Your Dog* by Chad Orzel. This book simplifies quantum physics by illustrating it through the many conversations he has with his dog. He also adds in the nitty-gritty quantum physics explanations. His book covers a wide range of topics that include the wave-particle duality, Heisenberg's uncertainty principle, the Copenhagen interpretation, quantum tunneling, quantum entanglement, quantum teleportation, and the misuse of quantum physics.

Another author that can help you learn more about quantum physics is Sean Carrol. He has a wide range of books that cover quantum physics as well as philosophical concerns that the advent of quantum physics brings about. Both of these authors are physicists in their own right, so the information can get complex. However, since I have laid a good foundation for you, I believe that you will have no problems understanding the content in their books.

There are also a variety of YouTube channels that delve into quantum physics. A good channel to subscribe to is Parth G. He is a physics graduate from Cambridge who loves to take complex science and make it easy to understand. He includes a lot of the math, so if you are interested in how the math of quantum physics works, then you can watch his videos. He makes the videos easy to understand that you do not have to be good at math to understand the formulas given.

I would advise against reading *Quantum Physics for Dummies* because it is way too advanced. However, I do think that you will benefit from looking through some high school textbooks that explain some of the classical physics. A good site to look at would be Siyavula's site. Aside from the textbooks provided on the site, they also have good illustrations that can really help you cement what you are learning.

Hopefully this book has given you a solid foundation of quantum physics to satiate your curious mind and inspire you to continue to be curious. Curiosity is a beautiful thing. Do not ever let the societal constructions of math and science tell you what you can and cannot learn. There have been many physical theorists that have done badly at school, and they still made great contributions to society. So go forth and learn more about quantum physics! Quantum physics is going to revolutionize the world as we know it. It is great to get a head start.

Best of luck, and thanks (again) for reading this book! If you found this book helpful, please leave a review to help others find it.

Bluesource And Friends

This book is brought to you by Bluesource And Friends, a happy book publishing company.

Our motto is **"Happiness Within Pages"**

We promise to deliver amazing value to readers with our books.

We also appreciate honest book reviews from our readers.

Connect with us on our Facebook page www.facebook.com/bluesourceandfriends and stay tuned to our latest book promotions and free giveaways.

Citations

Cresser, J. (2011, August 31). Quantum Physics Notes [Slides]. Unknown. https://physics.mq.edu.au/~jcresser/Phys304/Handouts/Quantum PhysicsNotes.pdf

Filmer, J. (2013, December 25). Sir Isaac Newton: Father of Modern Science. Futurism. https://futurism.com/sir-isaac-newton-father-of-modern-science-2

Hindu, T. (2017, August 6). What is Conformal Cyclic Cosmology? The Hindu. https://www.thehindu.com/sci-tech/science/what-is-conformal-cyclic-cosmology/article19436363.ece

How your Smartphone uses Quantum Mechanics. (n.d.). Quantum Technology. Retrieved July 6, 2020, from https://qt.eu/discover/applications-of-qt/how-your-smartphone-uses-quantum-mechanics/

Howell, E. (2017, March 30). Einstein's Theory of Special Relativity. Space. https://www.space.com/36273-theory-special-relativity.html#:%7E:text=The%20theory%20of%20special%20relativity,at%20the%20speed%20of%20light.

Johnston, H. (2018, June 25). Spatial overlap leads to useful quantum entanglement, say physicists –. Physics World. https://physicsworld.com/a/spatial-overlap-leads-to-useful-quantum-entanglement-say-physicists/

Orzel, C. (2009). How to Teach Quantum Physics to Your Dog. Scribner.

Physics: Newtonian Physics | Encyclopedia.com. (2020). Encyclopedia.Com. https://www.encyclopedia.com/science/science-magazines/physics-newtonian-physics#:%7E:text=Newtonian%20physics%2C%20also%20called%20Newtonian,Newton%20(1642%E2%80%931727).

Wikipedia contributors. (2020a, June 30). Physics. Wikipedia. https://en.wikipedia.org/wiki/Physics

Quantum Physics In Minutes

Wikipedia contributors. (2020b, July 6). Photoelectric effect. Wikipedia. https://en.wikipedia.org/wiki/Photoelectric_effect